my itty-bitty bio

Jerry C. Elliott High Eagle

CHERRY LAKE PRESS

Published in the United States of America by Cherry Lake Publishing Group
Ann Arbor, Michigan
www.cherrylakepublishing.com

Reading Adviser: Beth Walker Gambro, MS, Ed., Reading Consultant, Yorkville, IL
Book Designer: Jennifer Wahi
Illustrator: Jeff Bane

Photo Credits: © Danneellah/Shutterstock, 5; © Artsiom P/Shutterstock, 7; © alphaspirit.it/Shutterstock, 9; © Library of Congress/LOC No. 2004671908/Photo by Jack Orren Turner, 11; © Castleski/Shutterstock, 13, 22; © Public Domain/Wikimedia, 15, 23; © Roman Zaiets/Shutterstock, 17; © Rawpixel.com/Shutterstock, 19; © NASA/MSFC/Emmett Given, 21; Jeff Bane, Cover, 1, 6, 10, 14; Various frames throughout, Shutterstock images

Copyright © 2022 by Cherry Lake Publishing Group
All rights reserved. No part of this book may be reproduced or utilized in any form or by any means without written permission from the publisher.

Cherry Lake Press is an imprint of Cherry Lake Publishing Group.

Library of Congress Cataloging-in-Publication Data

Names: Thiele, June, author. | Bane, Jeff, 1957- illustrator.
Title: Jerry C. Elliott High Eagle / by June Thiele ; [illustrated by] Jeff Bane.
Description: Ann Arbor, Michigan : Cherry Lake Publishing, [2022] | Series: My itty-bitty bio | Audience: Grades K-1
Identifiers: LCCN 2021036545 (print) | LCCN 2021036546 (ebook) | ISBN 9781534198937 (hardcover) | ISBN 9781668900079 (paperback) | ISBN 9781668905838 (ebook) | ISBN 9781668901519 (pdf)
Subjects: LCSH: Elliott, Jerry C., 1943- | Physicists--Biography--Juvenile literature. | Physics--Juvenile literature.
Classification: LCC QC16.E55 T45 2022 (print) | LCC QC16.E55 (ebook) | DDC 530.092 [B]--dc23
LC record available at https://lccn.loc.gov/2021036545
LC ebook record available at https://lccn.loc.gov/2021036546

Printed in the United States of America
Corporate Graphics

table of contents

My Story 4

Timeline. 22

Glossary 24

Index 24

About the author: June Thiele writes and acts in Chicago where they live with their wife and child. June is Dena'ina Athabascan and Yup'ik, Indigenous cultures of Alaska. They try to get back home to Alaska as much as possible.

About the illustrator: Jeff Bane and his two business partners own a studio along the American River in Folsom, California, home of the 1849 Gold Rush. When Jeff's not sketching or illustrating for clients, he's either swimming or kayaking in the river to relax.

my story

I was born in 1943. I grew up in Oklahoma. I was raised by my single mother. I am **Osage-Cherokee**. My **native** elders named me "High Eagle" later in my life.

I wanted to work at **NASA**.
I dreamed of getting people to the Moon.

What do you want to be when you grow up?

Many people thought my dream was **unrealistic**. But I proved them wrong.

I studied **physics**. I worked hard. Scientist Albert Einstein and my mother inspired me.

Who inspires you?

My childhood dreams came true! I was the first **Indigenous** person to work at NASA. I helped land a man on the Moon.

I was awarded the **Presidential Medal of Freedom**.

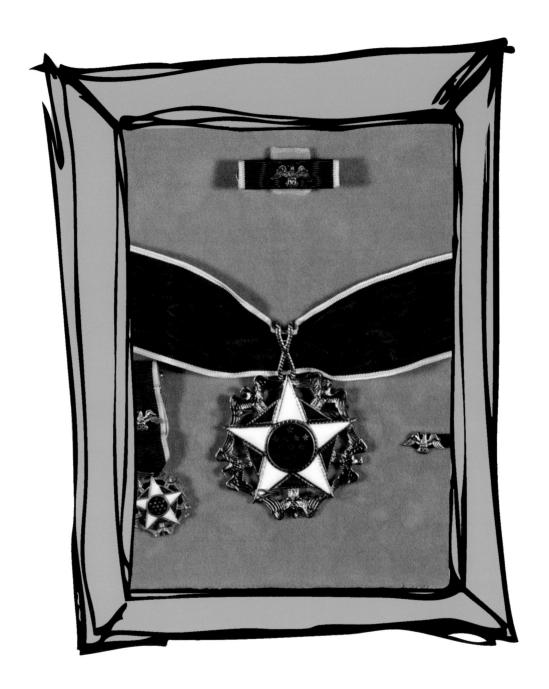

My friends and I started an organization. It helps Indigenous people enter the sciences.

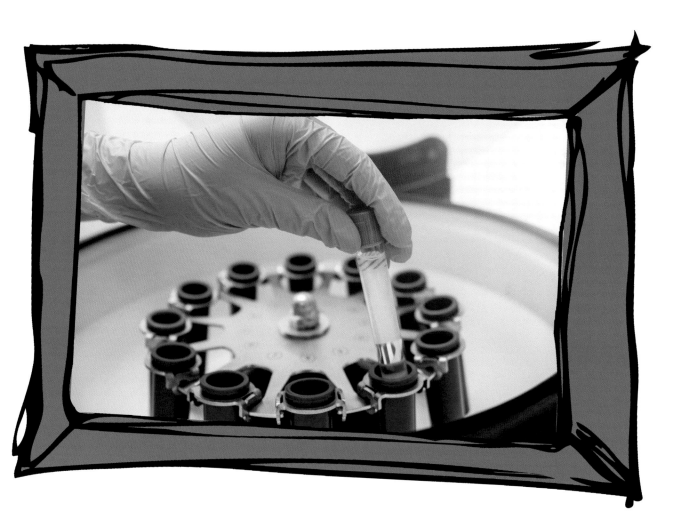

I started another organization.
It helps people with cancer.

I worked hard and dreamed big.
I help others dream big too.

What would you like to ask me?

timeline

1966

1940

Born
1943

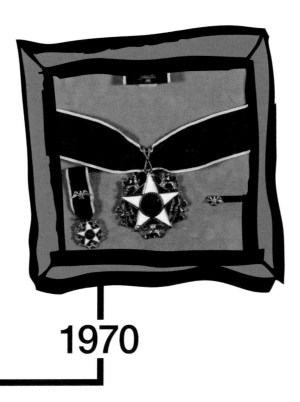

1970

2040

glossary

Indigenous (in-DIH-juh-nuhss) born or occurring naturally in a particular place; native

NASA (NAH-suh) National Aeronautics and Space Administration

native (NAY-tiv) born in a particular place

Osage-Cherokee (oh-SAYJ-CHAYR-uh-kee) a group of Indigenous people who lived in much of Oklahoma

physics (FIH-ziks) a science that deals with matter, energy, and motion

Presidential Medal of Freedom (preh-zuh-DEN-shuhl MEH-duhl UHV FREE-duhm) the highest civilian honor a person in the United States can receive

unrealistic (uhn-REE-uh-lih-stik) not likely to happen

index

Albert Einstein, 10

dream, 6, 8, 12, 20

Indigenous person, 12, 16

Moon, 6, 12

NASA, 6, 12

Oklahoma, 4
Osage-Cherokee, 4

physics, 10

science, 16